Table of Contents

    STEP-BY-STEP GUIDE 9

    1. Download and Install QGIS 9

    HOW TO INSTALL OR UPDATE QGIS 9

        DOWNLOAD THE INSTALLER 9

        INSTALLING QGIS WITH THE STANDALONE INSTALLER 10

        5.   Choose the installation location and shortcut options, then click 'Next'. 12

        6.   Click 'Install' and the installation will proceed to completion 12

        UPDATE OR INSTALL QGIS WITH THE OSGEO4W NETWORK INSTALLER ... 13

        4.   Select a URL to download from and click 'Next'. Any URL will work. I usually select http://download.osgeo.org. 14

    2. The QGIS Interface 15

        IMPORTANT PANELS 16

        Adding, Removing, and Moving QGIS Toolbars and Panels 17

    4. Map Navigation 19

    5. GIS Data Types 20

        VECTOR DATA 20

        RASTER DATA 21

        TABULAR DATA 21

    6. Add Data to QGIS 21

        Add Vector Data 21

        ADD RASTER DATA 25

        Layer Display Order 26

    7. Layer Properties 26

        Access Layer Properties 26

        VECTOR LAYER PROPERTIES 28

        RASTER LAYER PROPERTIES 28

    8. The Attribute Table 29

    9. Layer Symbology 31

        VECTOR LAYER SYMBOLOGY 31

        RASTER LAYER SYMBOLOGY 33

10. Create a New Shapefile 35
    Create the Vector Layer 35

11. Editing Vector Data 39
    Editing Feature Vertices 39

12. QGIS Toolboxes and Processing Tools 41

13. Create a Map Layout 43
    Create a Map Layout 43
    The QGIS Layout View 43
    Add Map to Layout 44
    Add Map Elements to the Layout 45
    Export Map 46

14. QGIS Plugins 46

A Beginner's Guide to QGIS

Getting to Know QGIS

Step-by-Step Approach

INTRODUCTION

A COMPLETE BEGINNER'S GUIDE

QGIS is a free and cross-platform Desktop Geographic Information System (GIS) application that supports viewing, editing, printing and analysis of geospatial data. QGIS works as Geographic Information System (GIS) software, allowing users to analyze and edit spatial information, in addition to composing and publishing graphical maps. QGIS supports raster, vector and mesh layers. Vector data is stored as point, line, or polygon features. Several raster image formats are supported and the software can geolocate the images. QGIS supports file formats, personal geodatabases, dxf, MapInfo, PostGIS and other industry standard formats. Web services, including Web Mapping Services and Web Feature Services, are also supported to allow the use of data from external sources. QGIS integrates with other open source GIS packages including PostGIS, GRASS GIS and Map Server.

Plugins written in Python or C++ extend the capabilities of QGIS. Plugins can geocode using the Google Geocoding API, perform geoprocessing functions similar to standard tools found in ArcGIS, and interface with databases PostgreSQL/PostGIS, SpatiaLite and MySQL. QGIS can also be used with SAGA GIS and Kosmo. Gary Sherman started developing Quantum GIS in early 2002, and it became an incubator project of the Open Source Geospatial Foundation in 2007. Version 1.0 was released in January 2009. , with the release of version 2.0, the official name was changed from Quantum. GIS to QGIS to avoid confusion as both names were used in parallel. Written primarily in C++, QGIS makes extensive use of the Qt library. In addition to Qt, required QGIS dependencies include GEOS and SQLite. GDAL, GRASS GIS, PostGIS and PostgreSQL are also recommended as they provide access to additional data formats.

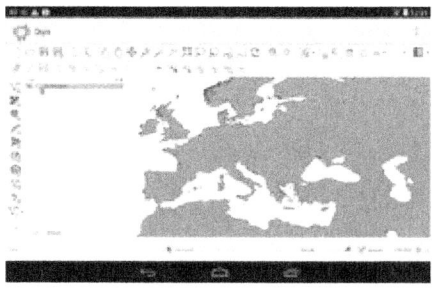

Screenshot of QGIS-Android in 2014.

As of 2017, QGIS was available for many operating systems including Mac OS X, Linux, Unix and Microsoft Windows. A mobile version of QGIS is being developed for Android since 2014. QGIS can also be used as a graphical user interface for GRASS. QGIS has a small installation footprint on the host file system compared to commercial GIS and generally requires less RAM and processing power; therefore, it can be used on older hardware or run concurrently with other applications where CPU power may be limited.

QGIS is maintained by volunteer developers who regularly release updates and bug fixes. Since 2012, the developers have translated QGIS into 48 languages and the application is used worldwide in academic and professional environments. Several companies provide support and feature development services.

KEY FEATURES

QGIS provides a continuously growing number of capabilities provided by core functions and plugins.

You can visualize, manage, edit, analyse data, and design maps.

Interoperability: support for numerous file formats and databases as well as web services.

Customizability: freedom to tailor the application to your needs, from custom data input forms to personalized user interfaces and workflows.

Extensibility: a C++ core and Python support provide the framework for everything from quick scripts to novel stand-alone applications based on the QGIS API.

COMMON FUNCTIONS

Geocode: To create points on a map from street addresses in spreadsheet form

Overlay: To superimpose two or more maps or layers in the same coordinate system, to show the relationships between them

Georeference: To align geographic data (map, layer, etc.) with a given coordinate system, allowing for overlays

Select by location (proximity analysis): To select features according to their relationship in space to other features

Select by attribute: To select features according to their properties (attributes), like querying a database

Buffer: To create a zone around a feature in units of distance or time

Network analysis: To find the distance from a feature travelled along a network (such as roads, public transit) rather than as the crow flies

Join: To attach fields from one table to those of another through an attribute or field common to both tables.

Viewshed analysis: To determine what areas are visible from a particular location.

COMMON DATA

Vector Data

Vector data represent discrete features, which could have names or attributes. It is often used in social science research to describe things with clear boundaries.

Vector features can be of three types:

1. point: a single pair of coordinates, like a cell phone tower or a crime scene

2. line: something that has length but not width; often a road or a river

3. polygon: an area with boundaries; often a political feature such as a state or county, but could also be a lake, building footprint, etc.

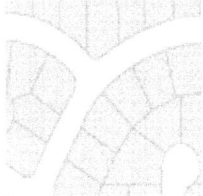

The most common file format for vector data is the shapefile (.shp). The shapefile is actually a collection of files: (.shp, .shx, .dbf).

Attribute Data

Attribute data are not really a separate kind of data, but rather the fields (numeric or text) that are attached to features.

ArcGIS can join together geographic data like shapefiles with tabular data in formats such as .xls, .mdb, .txt, or .csv, using the Join tool. The columns in the table become the attributes of the geographic feature to which it corresponds. The trick is finding a unique identifier common to the features in the shapefile and the rows in the table.

Raster Data

Raster data represent continuous features as cells (pixels) in a grid, much like image files in which each pixel has a location in space plus a meaningful value. Raster files are frequently used in scientific studies for phenomena like rainfall, elevation, or things that vary continuously in space.

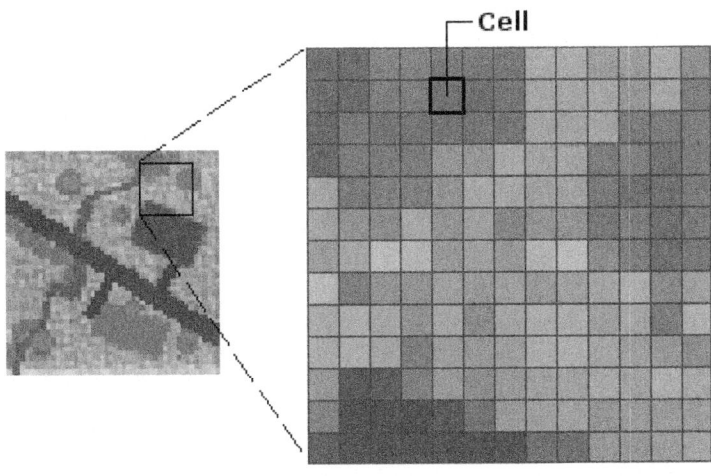

See **What is raster data?**

Common raster file formats include .img, .tif, and common image formats.

Where to find Data

A huge amount of spatial data is available free online, particularly through government agencies. International data may be among the most difficult to find.

1. **TIGER/Line Shapefiles**

    The U.S. Census Bureau makes available detailed shapefiles of census geographies (tracts, etc.) as well roads and other landmarks.

2. **Geo.data.gov**

    The federal government's one-stop data catalog.

3. **Pennsylvania Spatial Data Access**

    Official source for free shapefiles and other data formats specific to PA.

4. **National Atlas Data Download**

    From the U.S. Department of the Interior, a variety of shapefiles on environmental, socioeconomic, or agricultural topics.

5. **Global Administrative Boundaries**

    National and subnational boundary file for countries around the world.

6. **Geodata@Tufts**

    Browse the spatial data libraries of Tufts, Harvard, and MIT. Many files may not be available for download outside of those schools, but can give an idea of what exists.

    International GIS Data

STEP-BY-STEP GUIDE

1. Download and Install QGIS

QGIS can be installed for free on Windows, Mac, and Linux operating systems. The installation process is simple but it will take a little time for the installation files to download and be installed on your machine. This guide will walk you through QGIS installation, step-by-step.

HOW TO INSTALL OR UPDATE QGIS

QGIS is an amazingly powerful, free GIS software. In my opinion, it's the best free GIS software there is and even outperforms expensive, enterprise software in some instances (looking at you ESRI). QGIS is available for Windows, Mac, and Linux operating systems.

There are two different options for installing QGIS and the installation process can be a little different if you're installing for the first time or updating to a newer version. Using the workflow to update QGIS can save you time so that you're only installing the new software components. The installation and update processes are simple. I'll walk you through each one below.

DOWNLOAD THE INSTALLER

Go to qgis.org and navigate to the downloads page. On the QGIS downloads page, you will see a few different installer options.

For new users (or users that don't already have QGIS installed on their machine) the best option is the QGIS standalone installer (see image below). You can choose the latest release, which will be the most up-to-date and have the most features, or the long-term release, which is more stable but will have fewer features.

If you already have QGIS installed and are updating to a newer version you will want to use the OSGeo4W Network Installer (see image below). The network installer can also be used for a new installation.

The installation process is a little different using the standalone installer (for new installs) and the network installer (for updating an existing install) so I'll walk through the process for each.

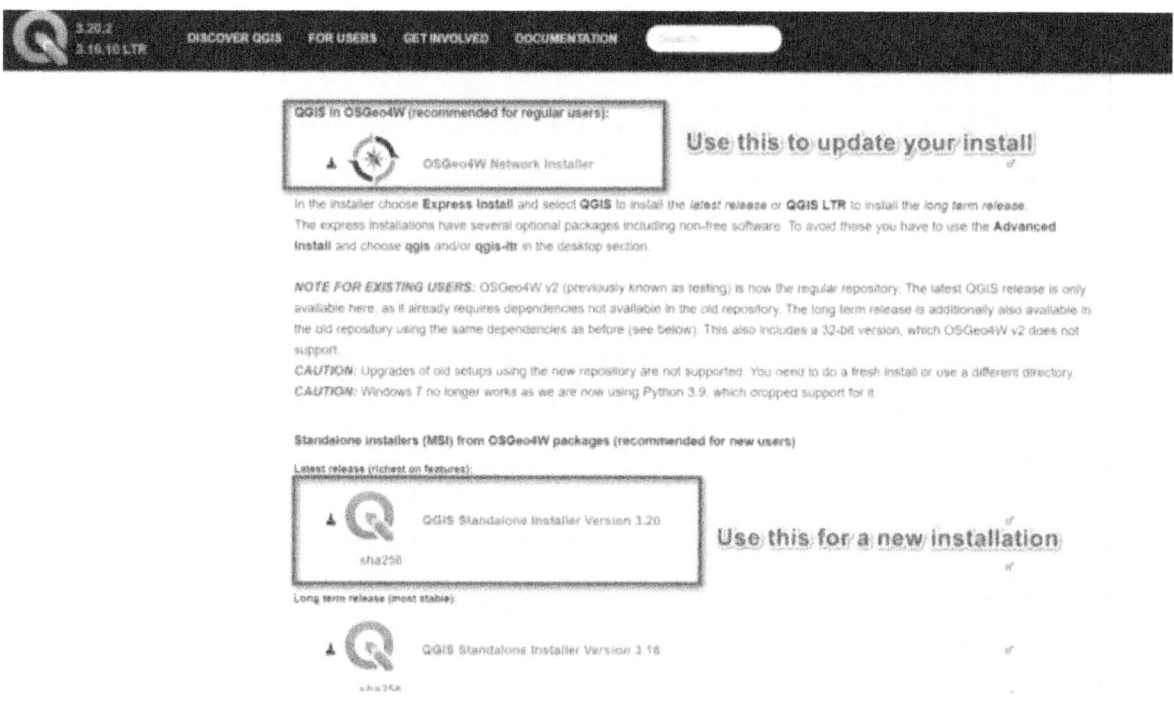

INSTALLING QGIS WITH THE STANDALONE INSTALLER

1. Download the standalone installer for your desired QGIS version. The installation file is quite large so it will probably take a little while to download.

2. Run the installation program once it has completed downloading.

3. The installer will open to a welcome message, click 'Next'.

4. Read and agree to the license terms and click 'Next'.

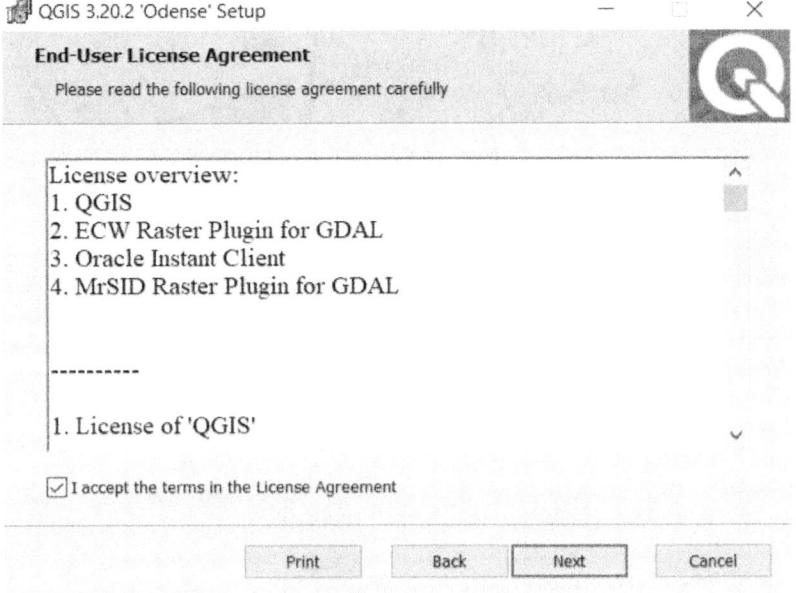

5. Choose the installation location and shortcut options, then click 'Next'.

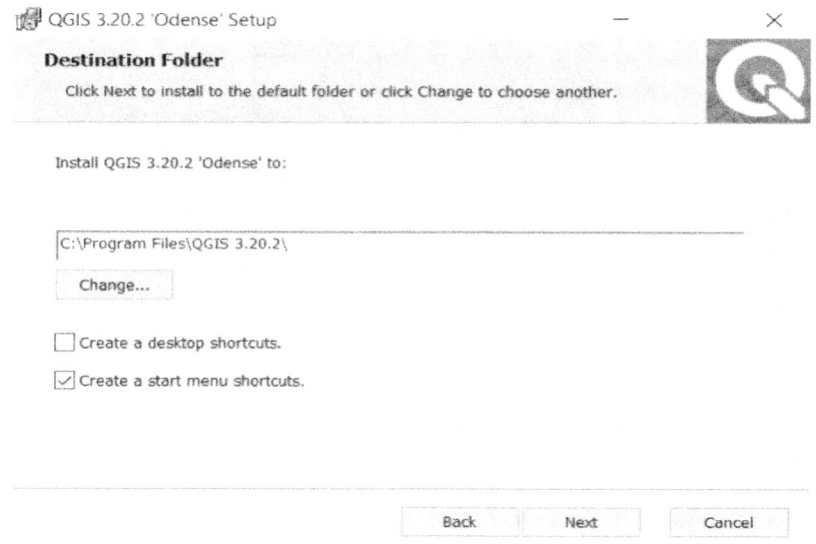

6. Click 'Install' and the installation will proceed to completion

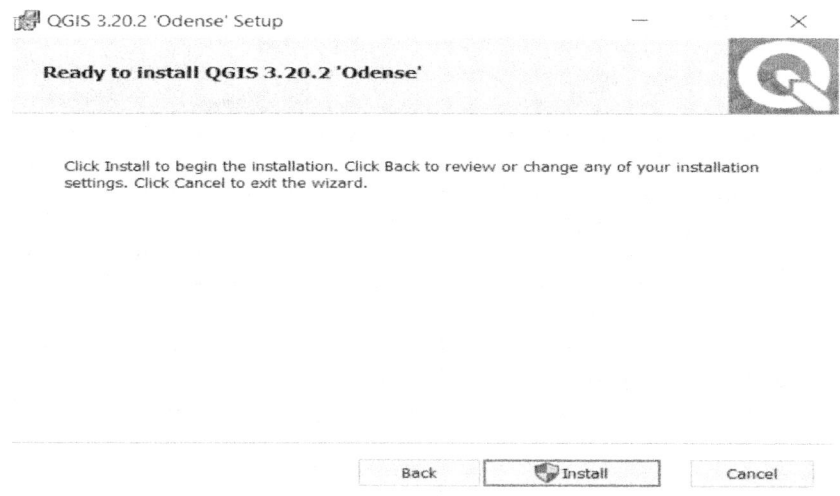

7. Once the installation is complete you can open QGIS and begin working!

UPDATE OR INSTALL QGIS WITH THE OSGEO4W NETWORK INSTALLER

1. Download the QGIS OSGeo4W Network Installer.

2. Double-click the downloaded installer to run the install program.

3. Select 'Express Install' and click 'Next'.

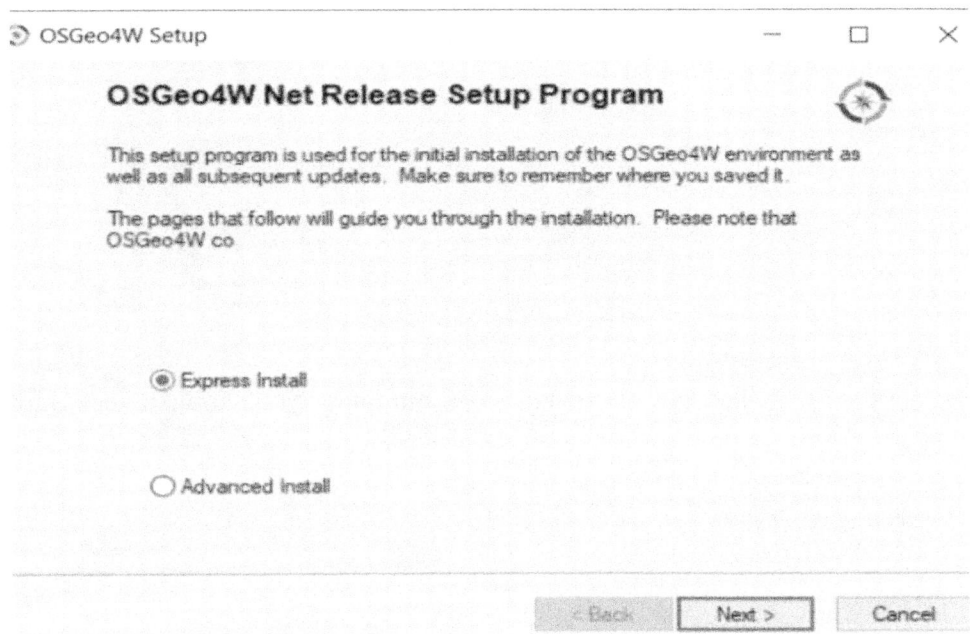

4. Select a URL to download from and click 'Next'. Any URL will work. I usually select http://download.osgeo.org.

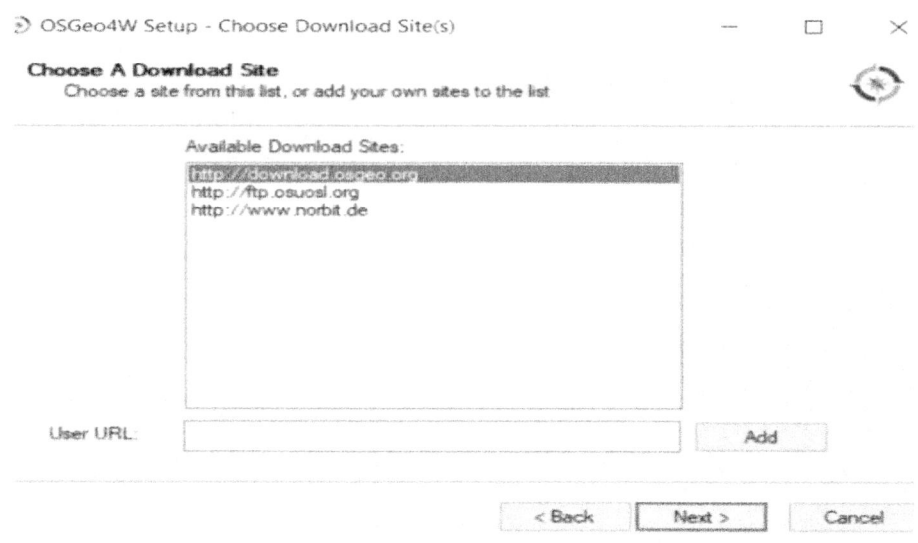

5. Select the QGIS packages to install. At a minimum you should select QGIS or QGIS

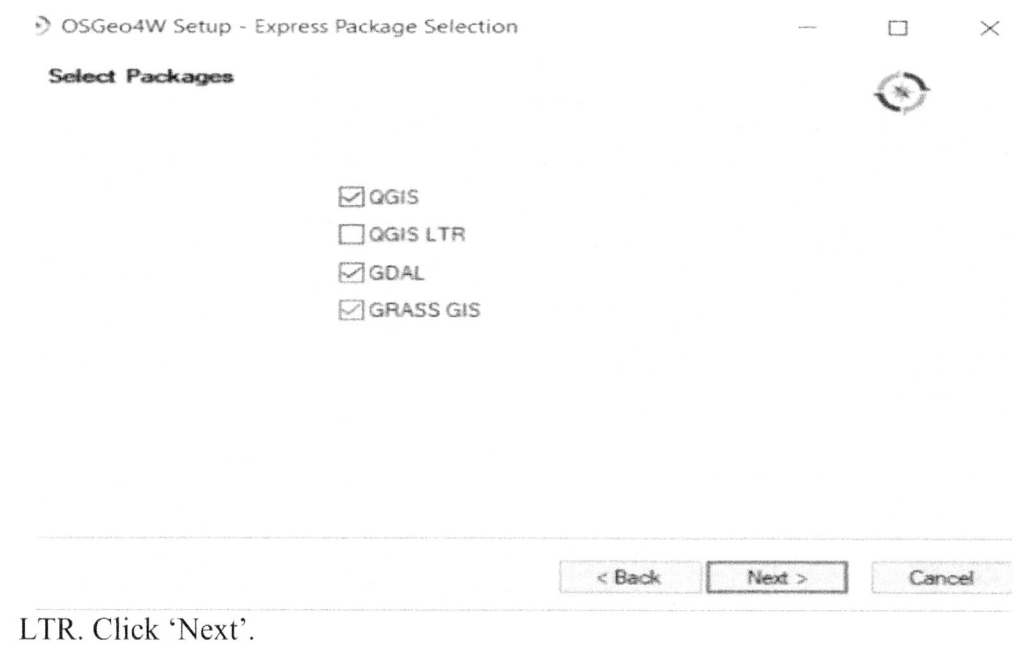

LTR. Click 'Next'.

6. When prompted, read and agree to the license terms for the different packages. After you agree to all the licensing the installation will begin. The installation will take quite awhile to complete. The more packages you select, the longer it takes. Installation time will partially depend on your internet connection

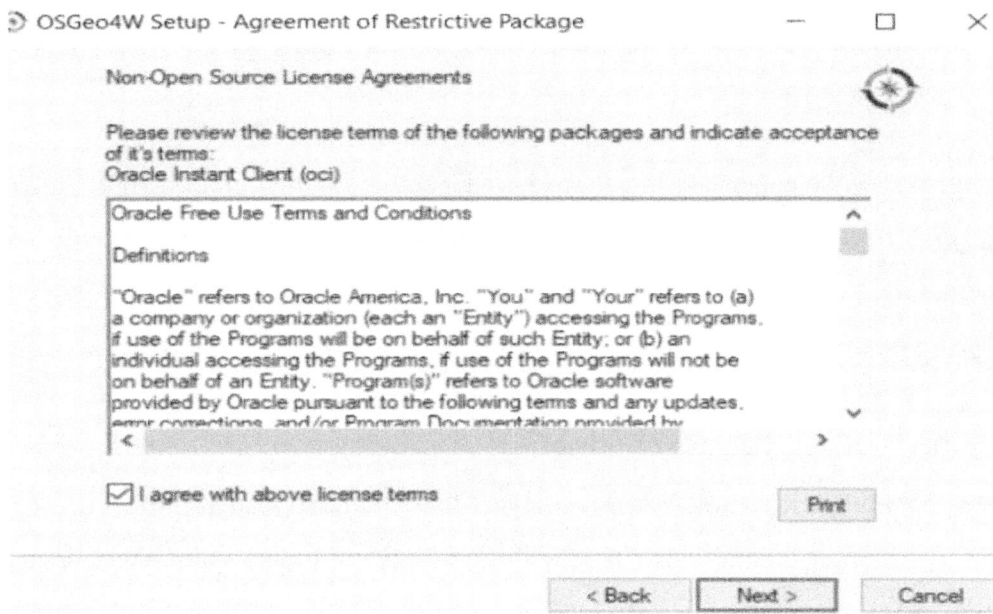

7. Click 'Finish' when the installation is complete.

8. Once the installation is complete, QGIS will be updated to the latest version. You can now open and use QGIS.

The QGIS installers and install process change slightly over time. These instructions are generally correct for QGIS installations and updates but some details may be slightly different for future QGIS versions. I'll try to keep this article updated with the most recent information.

2. The QGIS Interface

Once you have installed QGIS, run the program. This will take you to the main QGIS interface, which should look similar to the image below. Even if you have a newer QGIS version the layout and location of functions and tools should be almost identical. This tutorial was developed with QGIS 3.20.

The QGIS interface is composed of the main menu and a number of panels and toolbars. Panels and toolbars can be dragged around the interface. They can be docked to the interface edges or be free-floating.

IMPORTANT PANELS

There are two panels that will probably be open by default when you run QGIS for the first time. They are the Layers panel and the Browser panel. These are the panels you will probably use most frequently in QGIS.

In the image below the panels are in a tabbed configuration on the left side of the interface. Your panels may not be tabbed, as shown in the image, but you can create tabbed panels following the directions below. First I want to give you an overview of the Layers and Browser panels since we'll be using them frequently.

Layers Panel

The layers panel displays the different data layers that you have loaded into QGIS. Anytime you add a vector, raster, or tabular data file it will be displayed in the Layers panel. You can access information about these layers by right-clicking, or double-clicking, on a layer in this panel.

Note: Some operations will only be performed on a layer that is highlighted (or selected) in the Layers panel. You can highlight/select a layer by clicking on its name in the Layers panel.

Browser Panel

The Browser panel displays your computer's file structure and other data sources. From the Browser panel, you can navigate through directories and add files to the QGIS interface by clicking and dragging (we'll do that in just a minute).

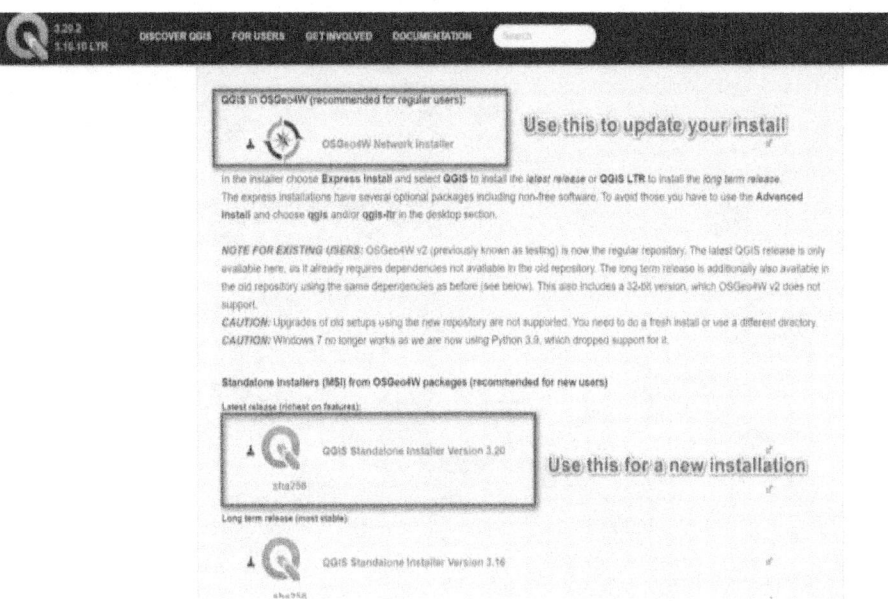

COMPONENTS OF THE QGIS INTERFACE.

Adding, Removing, and Moving QGIS Toolbars and Panels

If you find a toolbar or panel is missing it is easy to add it back to the interface. From the Main Menu select View. Then hover over the Panels or Toolbars option and you will see the available panels and toolbars. Select a panel or toolbar to add it to the interface (see image below).

You can remove a panel or toolbar by unchecking its box from the View menu.

QGIS toolbars and panels can be moved around the interface. To move a toolbar, click and hold on its far left side, then drag to the desired location. Toolbars will snap to the top and side of the QGIS interface.

Panels can also be moved. To move a panel, click and hold at the top of the panel, near its title, then drag to the desired location. Panels will snap to the top, side, and bottom of the QGIS interface. Blue boxes show the snapping location.

To create a tabbed panel (see image above) drag and hold one panel on top of another. A blue box will show where the panel will be snapped.

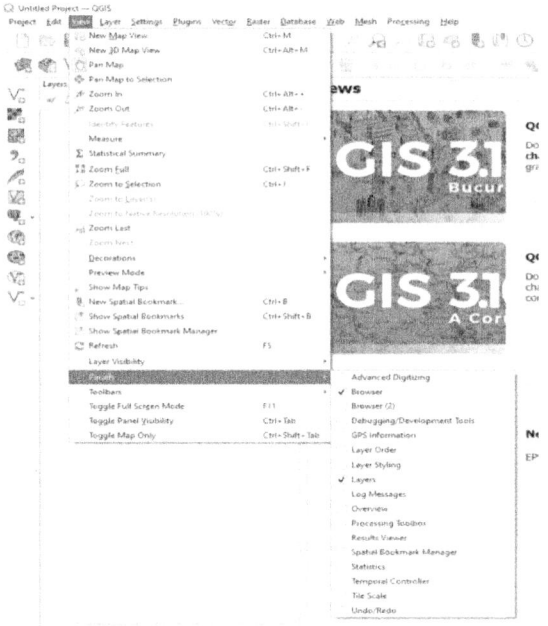

3. Add a Basemap to QGIS

Let's get started with an actual map. QGIS comes with Open Street Map available by default. You can access Open Street Map in the Browser panel (add the panel if you don't see it) under XYZ Tile (see image below).

To add Open Street Map, click and drag it from the Browser panel into the center map area of the interface (this may be the News or Project Templates area for new projects). Now you should have a basemap that covers the entire world.

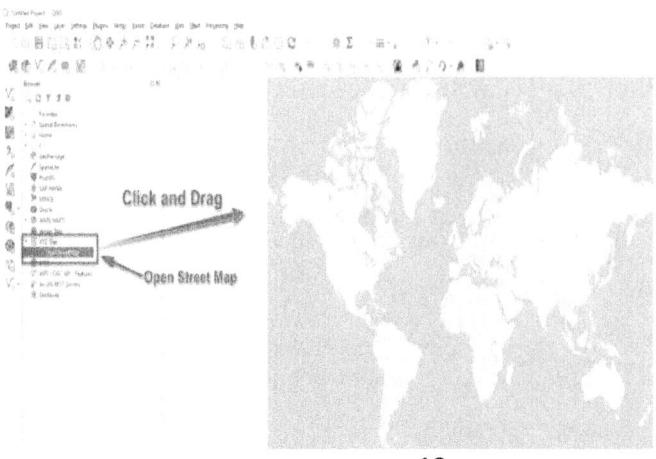

Note: Check out these tutorials to learn how to add other publicly available base maps like Google and other satellite imagery to QGIS.

4. Map Navigation

Tools on the Map Navigation Toolbar (see image below) and your mouse wheel are used to zoom and move around the map.

The QGIS Map Navigation Toolbar

Scrolling with a mouse will zoom in and out on the map. If you click and hold the mouse wheel and drag while the wheel is depressed you will pan the map.

In addition to the mouse wheel, you can use these basic tools from the Map Navigation Toolbar to navigate the map.

- **Pan** : Move the map around the screen. Once you click on pan, click and drag on the map to move it.

- **Zoom In** : Zoom in to an area on the map. A click will zoom in incrementally on the cursor location. Click and drag a rectangle to zoom in on an extent.

- **Zoom Out** : Zoom to a wider view. A click will zoom out incrementally from the curson location. Click and drag a rectangle to zoom out of an extent (less useful than click and drag with Zoom In).

- **Zoom Full** : Click this button to zoom to the full extent of the map. This will fit the whole map to the screen. It is useful if you zoom in or out too far.

- **Zoom to Layer** : Click this button to zoom to the full extent of the layer that is highlighted in the Layers panel.

Take some time to explore the map using the visualization tools. Zoom in to the area where you live or work and see if you can locate your address on the map.

5. GIS Data Types

It is important to understand the different types of spatial data and how they are represented. If you are new to GIS, these data types may be a little confusing at first. Don't worry. As you work with different data types you will become more familiar with them.

For now, just understand that there are different data types designed to represent different types of features and variables.

VECTOR DATA

Vector data show features that are represented by points, lines, or polygons. These data are discrete, which means they have specific boundaries and locations. Each vector feature (individual point, line, or polygon) has specific attributes associated with it.

For example, a point might represent the location of a fire hydrant. The attributes for that fire hydrant may include its color, street address, maximum water pressure, and most recent inspection date.

A line might represent a highway. Its attributes may include its length, speed limit, and maintenance condition.

A polygon may represent a building. Its attributes may include its size in square feet, street address, and maximum capacity.

In essence, vector data link a location (point, line, or polygon) to a row of information (attributes) that describe the location.

RASTER DATA

Raster data use a grid to represent things that are continuous in space. Each cell (or pixel) in the grid has a value. Raster data commonly represent things like precipitation, temperature, and elevation. These variables do not have discrete boundaries but change continuously.

For example, think of elevation. In any location the elevation changes in a continuous manner. As you walk up or downhill the elevation is constantly changing. A raster represents this change on the intervals of a grid.

TABULAR DATA

Think of tabular data like a table, or excel sheet. These data are not unique to GIS, but are commonly used to link outside information to vector data (and sometimes to rasters).

6. Add Data to QGIS

GIS data can be obtained from many different sources and can represent many different things. For this tutorial, I'm going to be using vector data from the US Census Bureau and raster data from the USGS 3D Elevation Program (3DEP). These data are freely available. Vector data are directly available from the census bureau and raster data can be downloaded from the USGS National Map.

Add Vector Data

I'll explain three ways you can add vector data to the map in QGIS. For this example, I'll be adding outlines for the 50 United States.

1. The Manage Layers Toolbar

Locate the 'Add Vector Layer' button on the

Manger Layers toolbar and click on it.

This will open the Data Source Manager window to load a vector layer. In the window, make sure 'File' is selected then click the ellipsis button to open up a file navigator. In the navigator, you can select the vector file you wish to add. After you select the file click the 'Add' button to add the data to the QGIS interface.

2. Drag and Drop From the Browser Panel

In my opinion, this is the easiest way to add a data layer to QGIS. Navigate to the file you want to add in the Browser panel (much like we did when we added OpenStreetMap). Once you have located the file, click and drag it into the map area of the QGIS interface.

That's all. It is that simple.

3. The Data Source Manager

Click the Data Source Manager button ![icon] on the Data Source Manager toolbar, or select Data Source Manager from the Layers drop-down in the main menu.

This will open a window like the one that appears when you click the Add Vector Layer button. Select the Vector panel on the left side, then add a vector layer following the directions above (for the Manage Layers Toolbar).

4. From the Layer Drop-Down on the Main Menu

Click the Layer option on the Main Menu. Hover over Add Layer from the drop-down list and select Add Vector Layer (see image below). This will open the Data Source Manager window and you can follow the directions from above to add a vector layer.

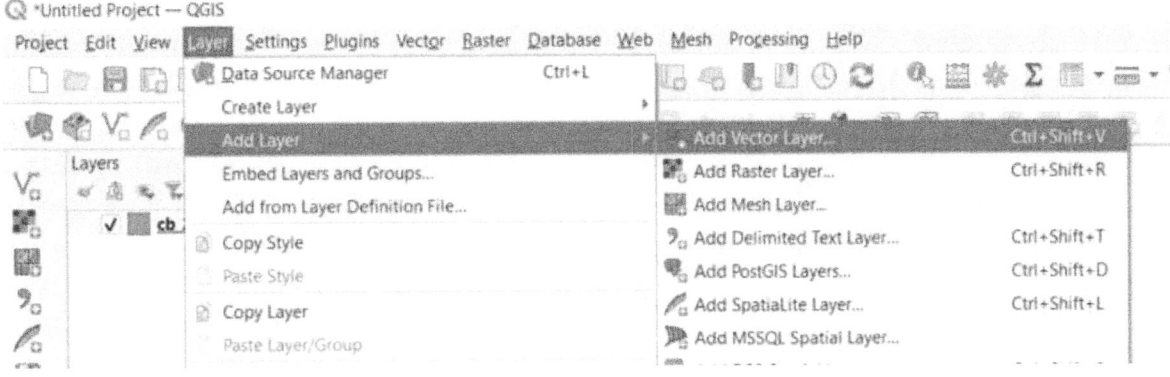

After You Add Data

You will probably get a pop-up notifying you about coordinate transformations (see below). Don't worry about this. Just keep the defaults and click OK. This is nothing you need to worry about right now.

It's also okay if you don't get the pop-up.

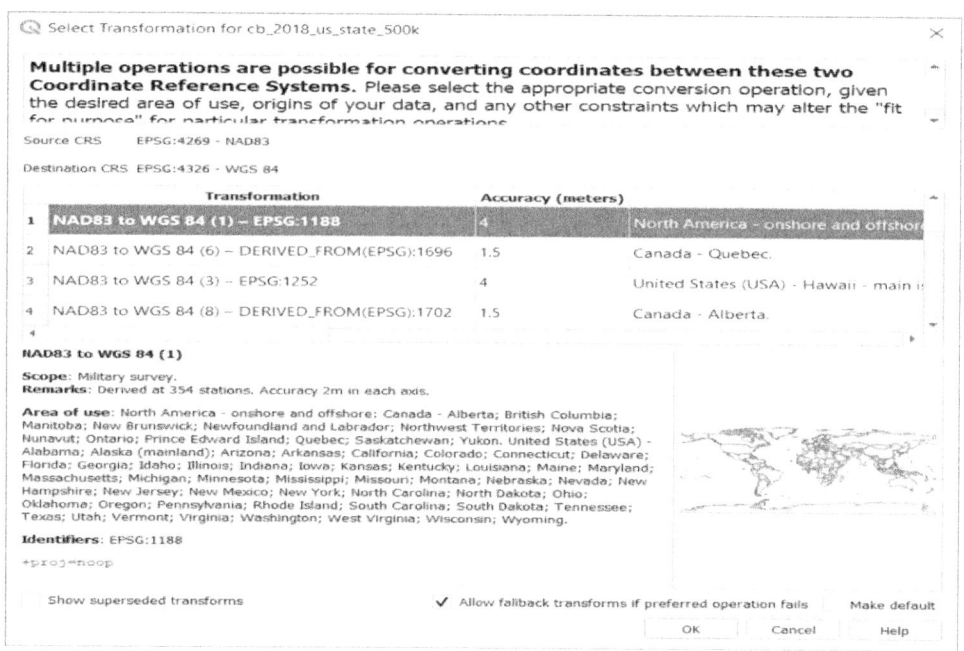

Once the data are successfully added to the map interface you will see the outlines show up on the map. An entry will also be added to the Layers panel (see below). The image below shows the result of adding the outlines of the United States to QGIS

ADD RASTER DATA

Raster data can be added using the same four methods I described to add vector data. There are a couple of differences so I'll explain those here.

1. The Manage Layers Toolbar

From the Manage Layers toolbar click the 'Add Raster Layer' button   This will open a Data Source Manager window that is very similar to the window opened to add vector data. Now just select the file you want to add and click OK to add the raster layer.

2. Drag and Drop from the Browser Panel

In the Browser Panel, navigate to the raster file you wish to add, then drag and drop it into the QGIS map area.

3. The Data Source Manager

Follow the same method to open the Data Source Manager as if you were adding vector data. Select the Raster tab on the left side of the window. Then select the raster layer you wish to add and click OK.

4. From the Layer Drop-Down on the Main Menu

Click the Layer option on the Main Menu. Hover over Add Layer from the drop-down list and select Add Raster Layer (see image below). This will open the Data Source Manager window and you can follow the directions from above to add a raster layer.

After You Add Data

The image below shows the raster layer that I added to the map. It is a digital elevation model for an area in Idaho. Notice that there are now two layers showing in the Layers panel. Next, we'll talk about the properties of vector and raster layers.

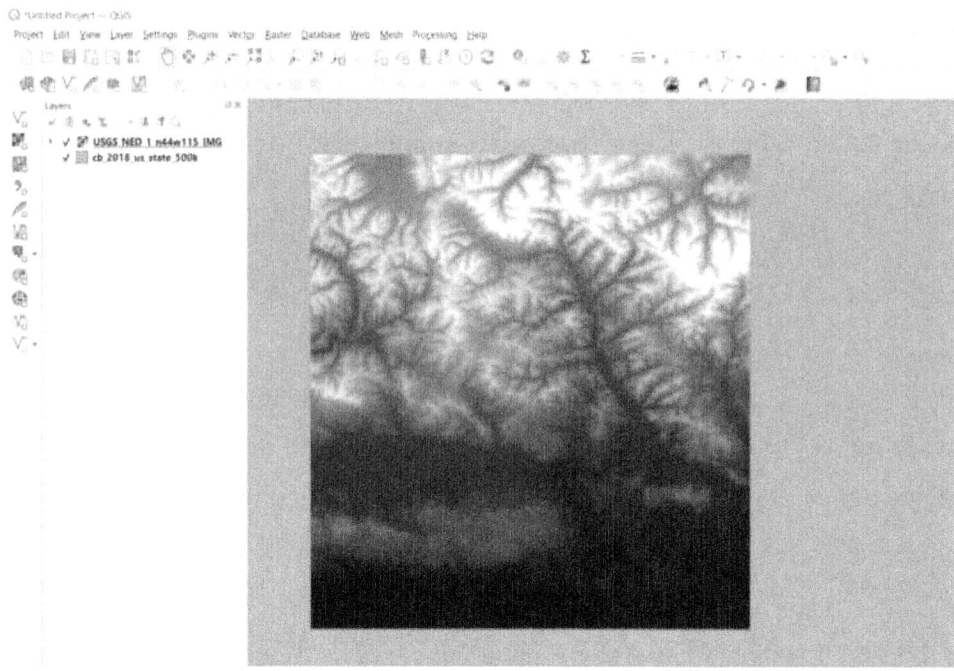

Layer Display Order

You can change which layer appears on top by dragging layers above or below others in the layer panel. This can also be done by single-clicking a layer so that it is highlighted in the Layers panel, then using the up and down arrows (at the top of the Layers panel) to change to the order in which it appears.

If you've added a layer that isn't showing, make sure it's not hidden below another layer.

7. Layer Properties

Every raster and vector layer has a number of properties associated with it. These properties display important information about the layer, let you change layer settings, let you change the way the layer looks, and let you add information to a layer.

There are numerous properties for each layer. In this tutorial, I will only go over the basics.

Access Layer Properties

To access the properties for any layer in QGIS right-click on the layer name in the Layers panel and select Properties. You can also double-click on the layer name in the Layers panel.

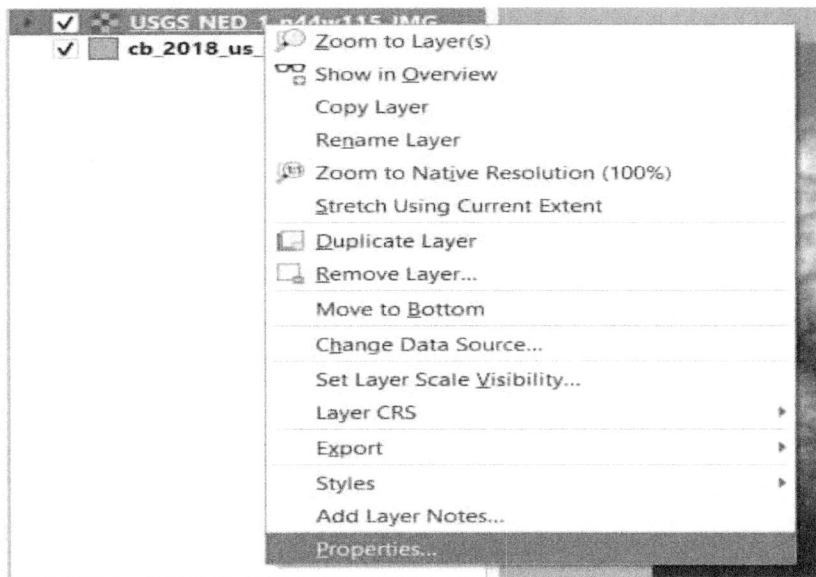

This opens a new window with several tabs on the right side that give access to layer information and settings. Layer properties will be different between vector and raster layers and can even differ between different file formats and data types. The image below shows the properties window for a vector layer of the states of the USA.

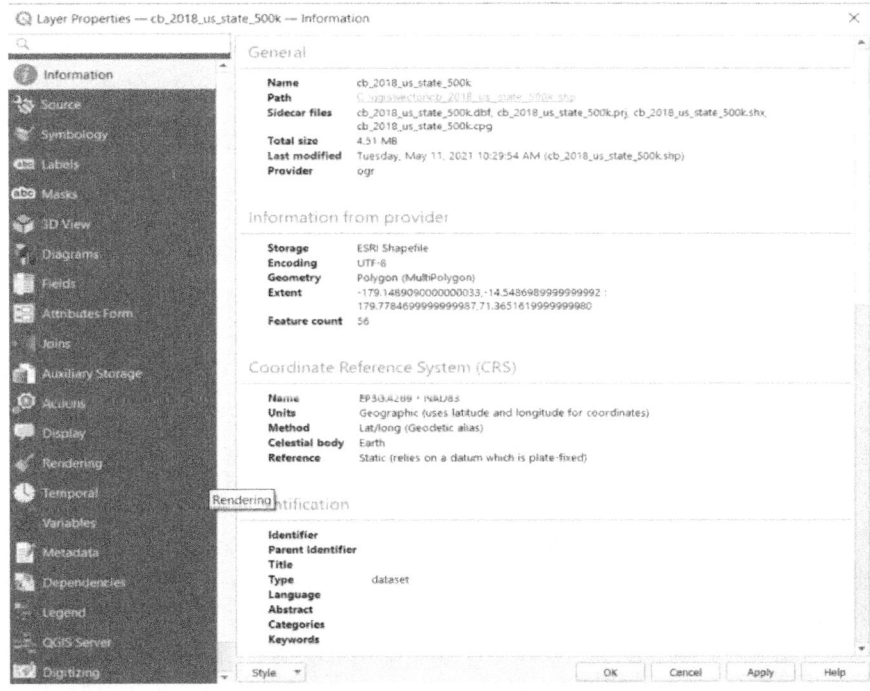

VECTOR LAYER PROPERTIES

Information and settings related to vector layers can be accessed from the properties tab. I won't go into all these things in-depth in this tutorial but will give you a brief overview.

The Information tab in the properties window displays basic information including the layer's coordinate reference system, file location, spatial extent, field names and data types, and other metadata specified by the file creator.

Symbology and Labels are two of the tabs you will probably use most in the layer properties window. I'll cover symbology in greater depth later in this tutorial. Selecting the appropriate tab will give you access to options and properties to adjust symbology and labels.

Field and attribute information and settings can also be accessed in the properties window.

Vector layers can be joined (linked) to other tabular data sources using the Join tab on the QGIS properties bar. Simply click on the Join tab, click the plus button to add a join, then specify the fields that link the vector layer and join layer. Once complete, a record will be added to the list of joins on the join tab.

RASTER LAYER PROPERTIES

Fewer properties are available for raster layers. In the information tab, you will be able to view the coordinate reference system, spatial extent, file location, and metadata associated with the raster layer.

The symbology tab gives access to change the symbology and display settings of a raster layer.

The properties window also gives options to adjust the transparency of a raster layer and to view a histogram of all values for a raster.

8. The Attribute Table

Each vector layer has an associated attribute table. The attribute table provides information that is specific to each geographic feature. Most raster layers do not have attribute tables, though raster attribute tables are created in some instances.

To access the attribute table of a vector layer in QGIS, right-click the vector layer name in the Table of Contents panel. Then select Open Attribute Table. This will open a new window that displays the feature attributes as a table with additional tools and options for selecting and interacting with the data.

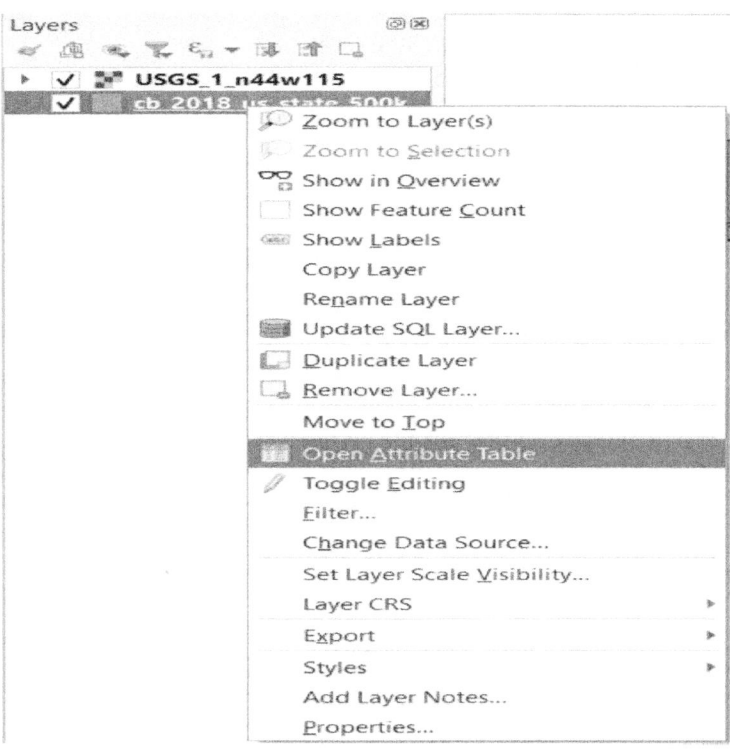

In the attribute table for states of the USA you will see that each feature has an associated name, abbreviation, and area, along with other attributes. A specific feature can be selected by clicking the index number on the far right side of the attribute table. The selected row in the attribute table will be highlighted in blue and the corresponding feature will be highlighted in the selection color (cyan in the image below) on the map.

Multiple features can be selected by holding down the control key while clicking the index number of multiple rows.

A subset of the attribute table can be displayed by selecting the features to show in the bottom right corner of the attribute table (see image below). This feature will allow you to show in the attribute table only selected features only features that are currently visible in the map, edited or new features, or to create a custom filter

To clear the selection, use the clear selection button on the Attribute Table Toolbar or the Selection Toolbar.

Values in the attribute table can be changed by editing a vector layer. Creating and editing vector layers is discussed later on in this tutorial.

9. Layer Symbology

There are two primary ways to access layer symbology options in QGIS. First, from a layer's Properties window. The section on layer properties describes where to find the symbology options. Second, from the Layer Styling panel. You can add this panel to the QGIS interface as described above in the QGIS Interface section.

Symbology options will be different for raster and vector layers, so I'll describe each one separately here.

I'll demonstrate symbology using the Layer Styling panel, but all these options can also be accessed from the Symbology tab in the layer properties window.

VECTOR LAYER SYMBOLOGY

To start, make sure the appropriate layer is selected in the Layer Styling panel. Below the layer name, you will notice an option to select the type of symbology to apply. For this tutorial, we'll just focus on Single Symbol symbology. The other symbology options for vector layers are:

- **Categorized**. Categorized symbology displays features based on an a categorical variable. An example of this is land cover type where a different symbol is used for features that describe forest, grassland, bare earth, water, and urban areas.

- **Graduated**. Graduated symbology displays features with a color and/or size gradient based on continuous values. An example of this is displaying cities by population where larger cities are represented by a larger symbol or darker color than smaller cities.

- **Rule-based**. Rule-based symbology gives the user more control over how features are symbolized based on a suite of variables.

After you understand the basics of symbology play around with these more advanced options. You will use Single Symbol, Categorized, and Graduated symbology most often.

For polygons, the fill color, fill type, stroke color, and stroke weight can be changed by clicking Simple Fill. When you click on the drop-down arrow next to a color, a new window will open with a color wheel so you can choose any color.

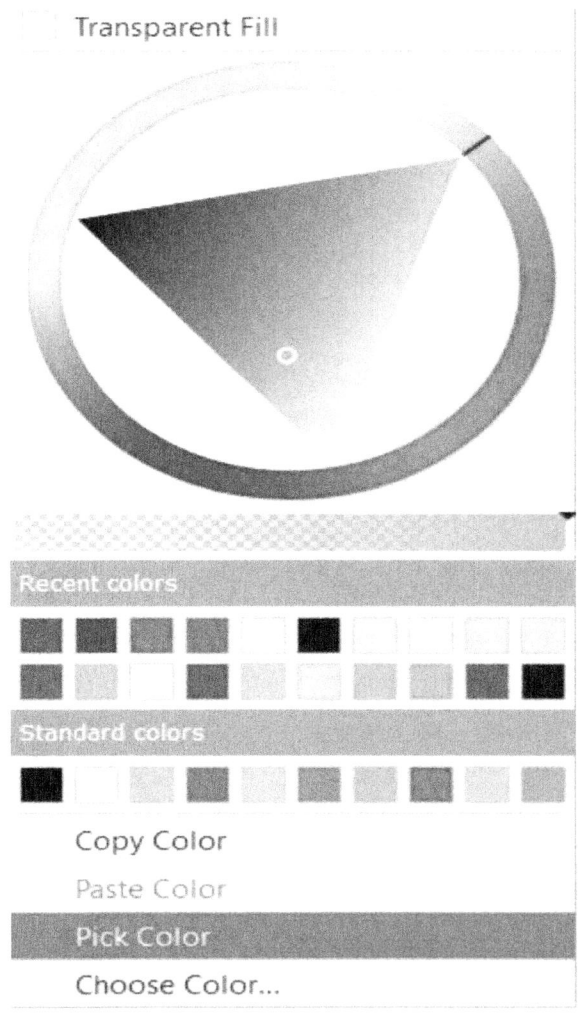

Here I've changed the default symbology to a blue stroke with a yellow diagonal fill outline.

When you use the Layer Styling panel your changes will apply automatically if the 'Live update' box is checked. If you are changing symbology from the properties window you will need to click 'Apply' or 'Okay' before the changes are applied.

The process for symbolizing points and lines is nearly identical. There will just be different options for the point and line styles.

RASTER LAYER SYMBOLOGY

Make sure a raster layer is selected in the Layer Styling panel. Then select the raster symbology method you wish to apply. Here I will use Singleband pseudocolor. The available symbology options for rasters are:

- **Multiband color** – displays raster data with multiple bands where different bands are shown in red, green, or blue. This is how aerial imagery and multiband imagery are displayed in true color and false color composite.

- **Paletted/Unique values** – displays each unique value in a raster with a different color or symbol. This is most often used for raster with few unique values such as land cover type or presence/absence rasters (not good for data like elevation and climate).

- **Singleband gray** – default raster symbology in QGIS that displays raster data on a back to white scale.

- **Singleband pseuodcolor** – displays continuous raster data along a specified color gradient (or color ramp). This is commonly used form many types of conintuous raster data including elevation and climate.

Now you can select the color ramp of your choice to change the raster's symbology. Here I've selected the 'Turbo' color ramp.

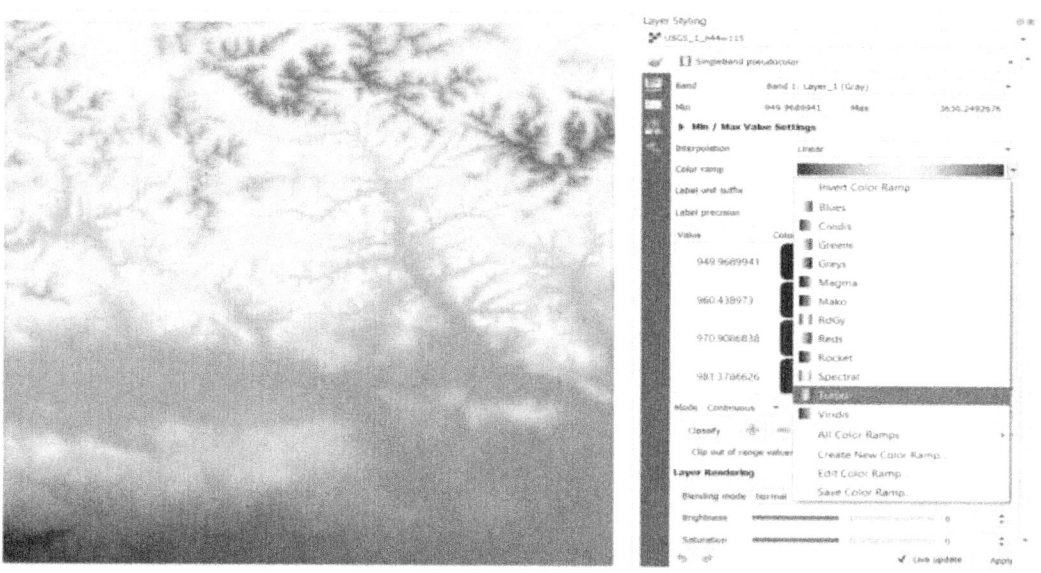

The 'Mode' can be changed to adjust how the color ramp is stretched. The default mode is 'Continuous' but 'Equal interval' and 'Quantile' are also available. Play around with these modes to see how it changes the symbology of your raster layer.

The interpolation method can also be changed along with mode to get different symbology effects. In the image below I've used discrete interpolation with the equal interval mode to create five elevation groups where elevations between certain values all have the same symbology.

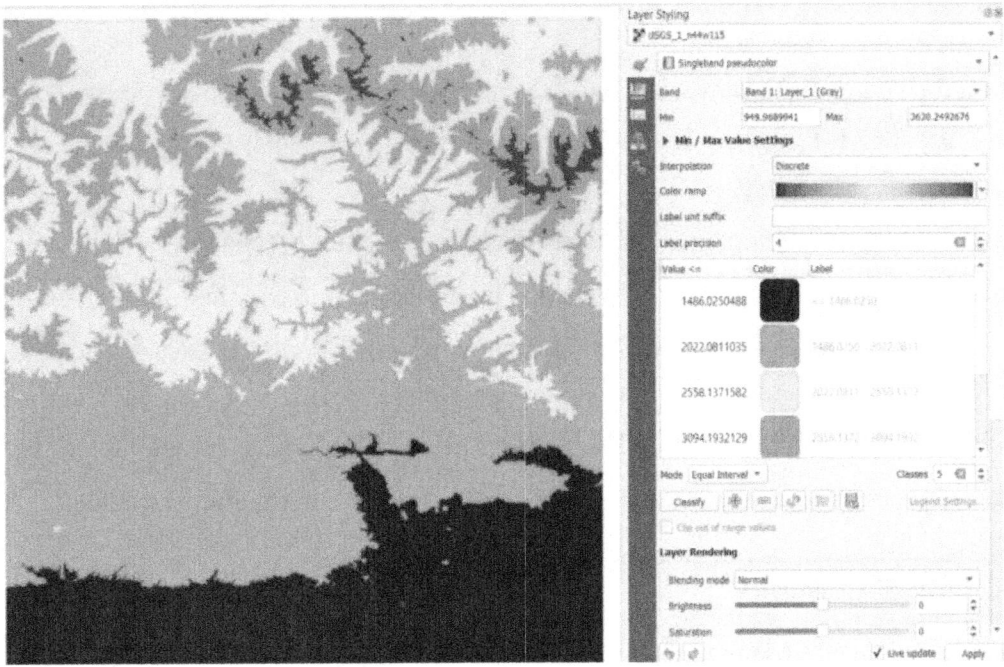

This is a very brief overview of how to change symbology. I encourage you to experiment with different symbology options.

10. Create a New Shapefile

Create the Vector Layer

To create a new shapefile in QGIS select 'New Shapefile Layer' from the 'Manage Layers Toolbar' or select Layer > Create Layer > New Shapefile Layer from the Main Menu.

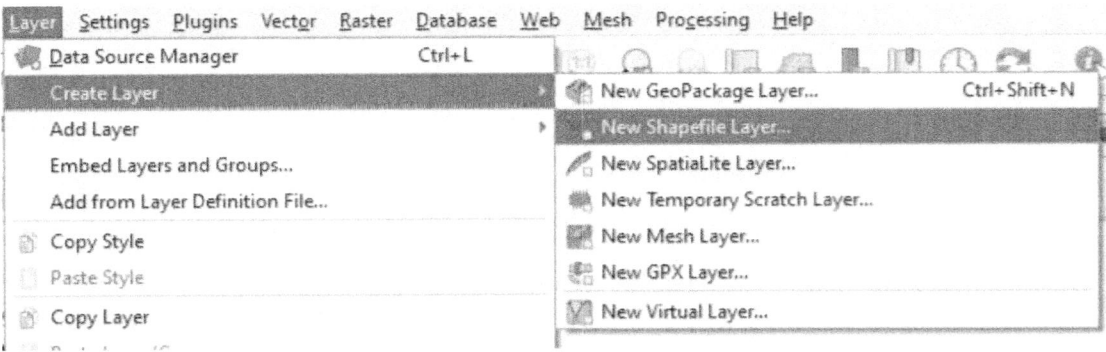

This will open a new window where you can specify a file name, geometry type, coordinate reference system, and create fields for the new layer. I've chosen the default CRS (WGS 84).

This is fine for you to do to start with, but depending on your end purpose you may want to use a projected CRS (like UTM or Albers).

To add a field, type the field name, select the field length (and precision, the number of decimal points, if it's a decminal), and click add field. This will add it to the field list. If you don't know what your fields will be it's okay you will be able to create more fields after the layer is created.

By default an the 'id' field is created. I've created three other fields as examples to showcase the different data types. You can see how I've set up my shapefile in the image below.

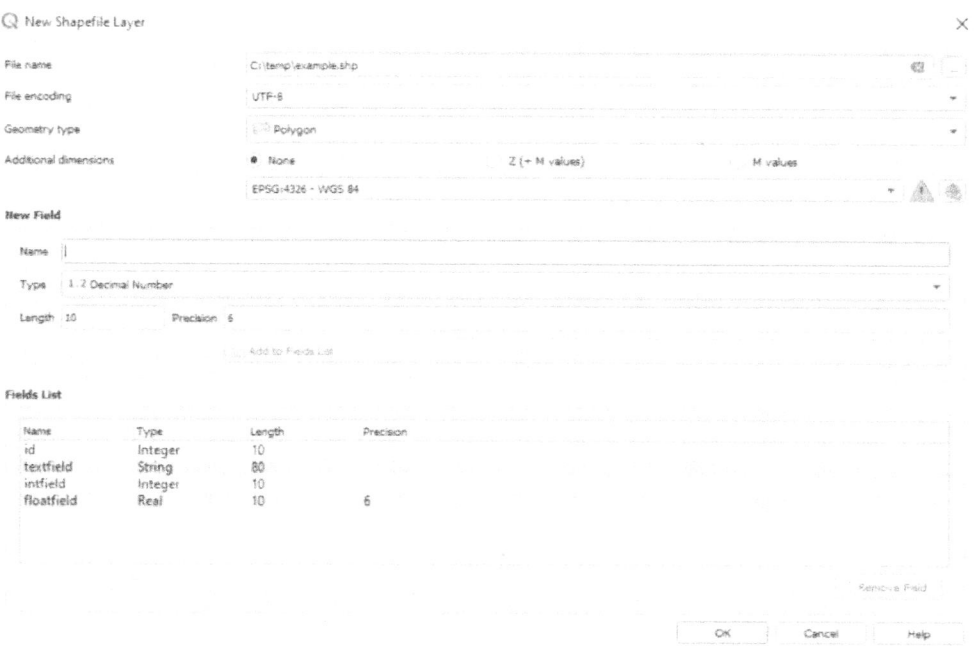

When everything is set up, click 'OK' and your new shapefile will be created and added to the Layers panel.

Create Features

If you open the attribute table of the new layer, you'll notice that it there are not any features. That's because we haven't added any features to the layer yet.

To add a new feature to the layer we need to turn editing on for the layer. First, make sure the layer you want to add features to is selected in the Layers panel. Then turn editing on by clicking the 'Toggle Editing' button on the Digitizing Toolbar. You can also select Layer > Toggle Editing from the Main Menu. You should now see the editing icon next to the layer name in the Layers panel, as shown below.

If the edit icon appears next the wrong layer, click on that layer to highlight it in the Layers panel the click the toggle editing button to turn off editing for the layer. Then highlight the correct layer and toggle editing on.

Once editing is toggled on more options will become available on the digitizing toolbar. To add a feature select the 'Add Polygon' button on the digitizing toolbar. This icon will appear slightly different for adding point or line features and depends on the geometry type you selected when you created the shapefile.

Once a vector layer or file is created the geometry type cannot be changed.

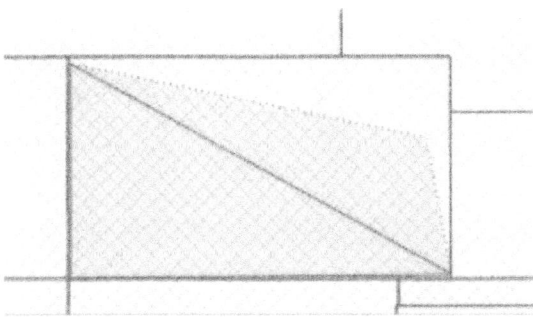

After you click the Add Polygon (or other geometry) button the cursor will change to crosshairs. This indicates when you click it will create new vertices for a feature. Left-click to add a vertex for a new feature and keep clicking to add additional vertices. As you add vertices the outline and area of the polygon will be updated (see image below).

When you are finished creating the outline of a feature, right-click to finish the feature. This will open a window for you to add the values for the feature.

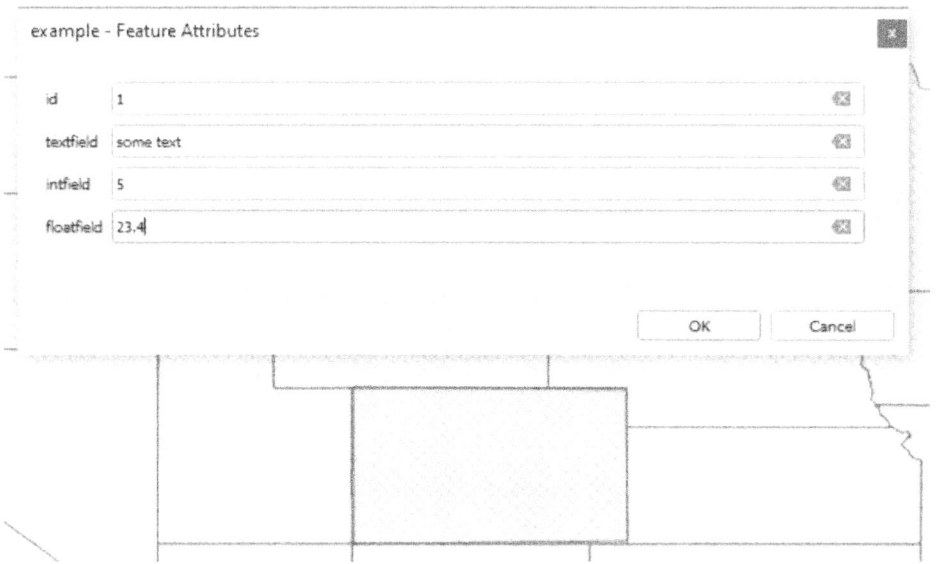

Once you add the values and click okay, the feature will be created and added to the attribute table. While editing mode is toggled on you can edit values directly in the attribute table.

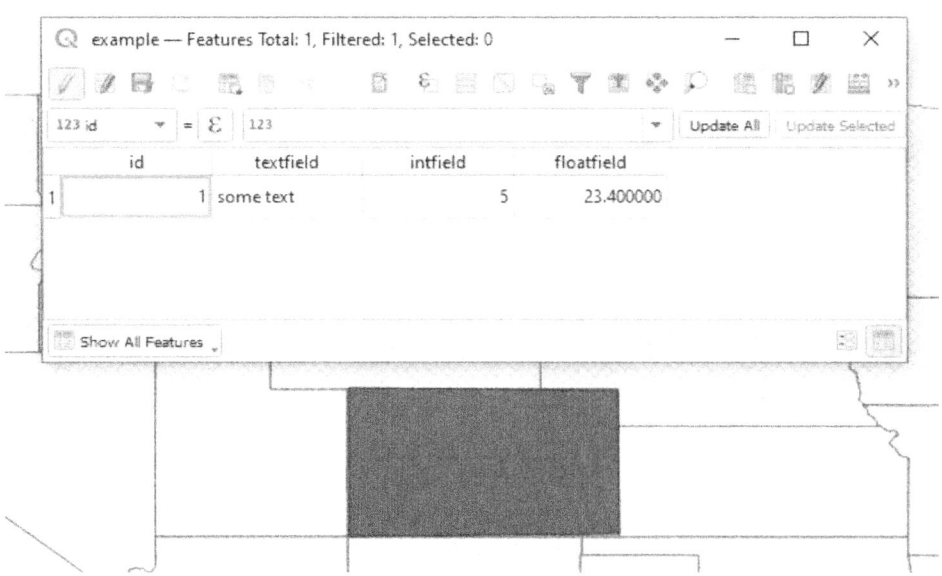

Once you have finished creating features click the 'Save edits' button on the Digitizing Toolbar then toggle editing off. Alternatively, you can just toggle editing off and choose the option to save edits.

11. Editing Vector Data

To edit feature shapes and attributes in QGIS toggle editing on by clicking the 'Toggle Editing' button. The pencil icon should appear next to the proper layer in the Layers panel.

Editing Feature Vertices

You can change the shape of a polygon or line or the location of a point with the

Vertex Tool

When you click on the vertex tool it will open a new window with a message that instructs you to click on a feature to view and edit its vertices.

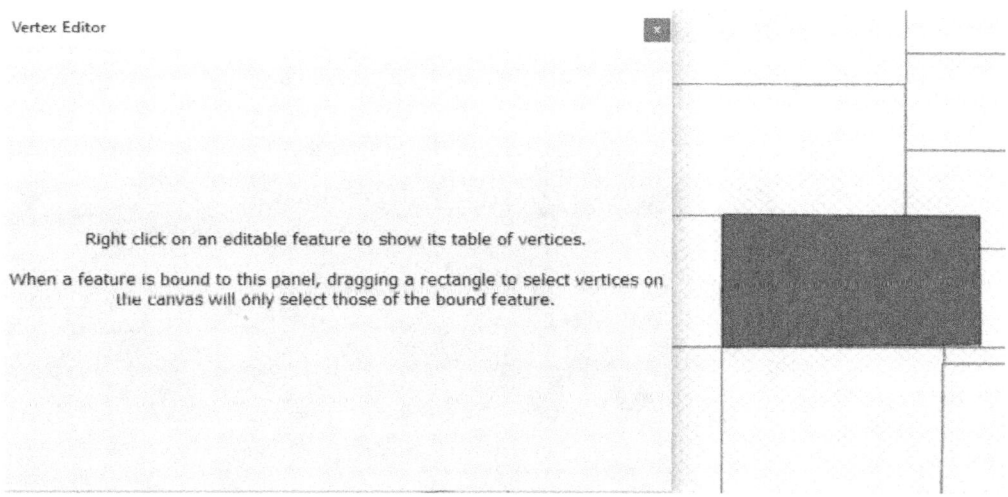

After clicking on a feature you will see a list of vertices.

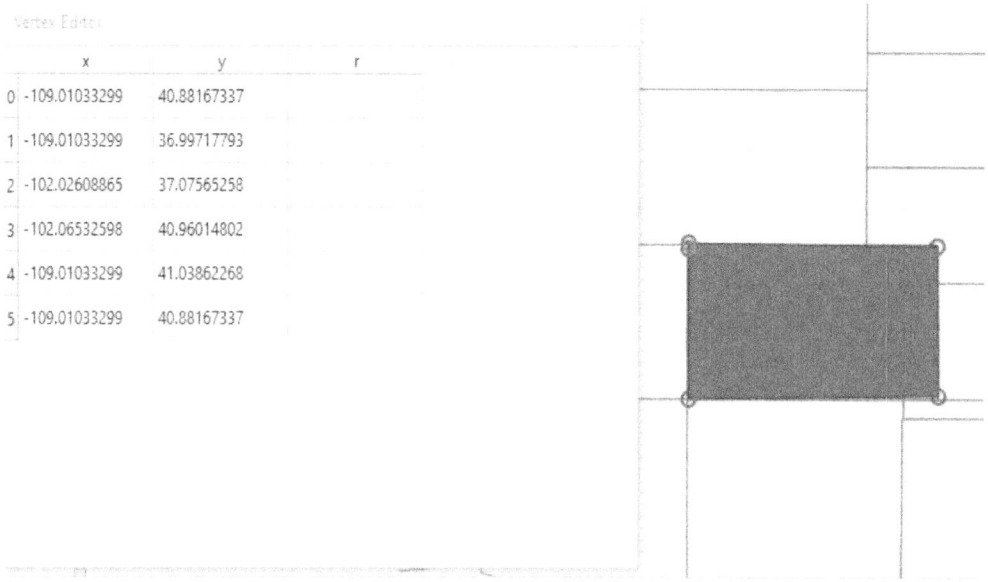

To move a vertex left click on it. You will now see a dotted line showing what the new feature will look like if you move the vertex to the location of the cursor. When the cursor is in the location you desire left click to move the vertex.

You can add a new vertex by hovering the cursor near the midpoint of an existing line or polygon segment. A plus sign (+) will appear. Click the plus sign to add a new vertex, then click again at the location you wish to place the vertex.

As you move and add vertices they will be added to the vertex table in the Vertex Tool window. You can also edit the vertex coordinates directly in the table.

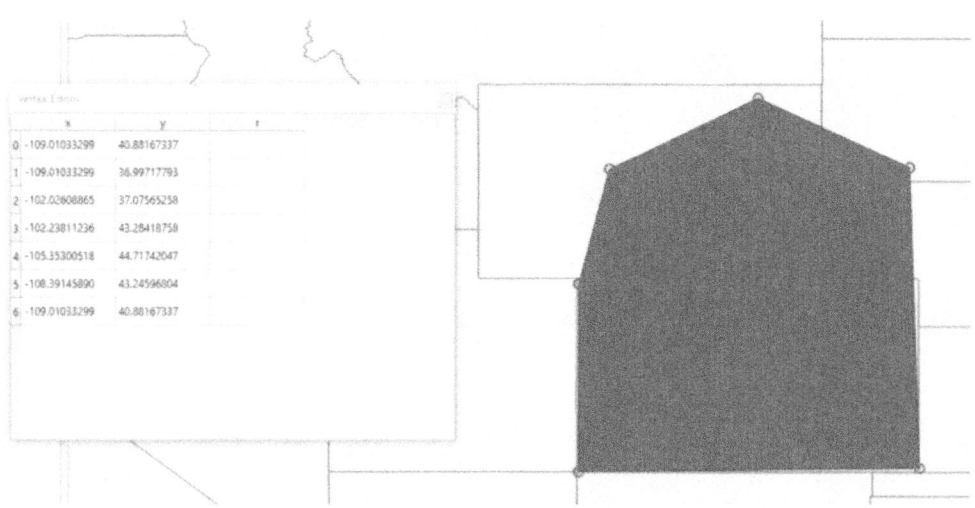

Editing Feature Attributes

To edit feature attributes, open the attribute table. If editing mode is not on, you can toggle it on directly from the attribute table with the toggle editing button.

Once editing is on you can make changes to the feature attributes in the attribute table. You can also use the field tools to add and remove columns (fields).

12. QGIS Toolboxes and Processing Tools

QGIS has many built-in functions and tools for analyzing and interacting with geospatial data that are located in the Processing Toolbox. The processing toolbox can be opened by clicking on its icon in the Attributes toolbar or by selecting Processing > Toolbox from the Main Menu.

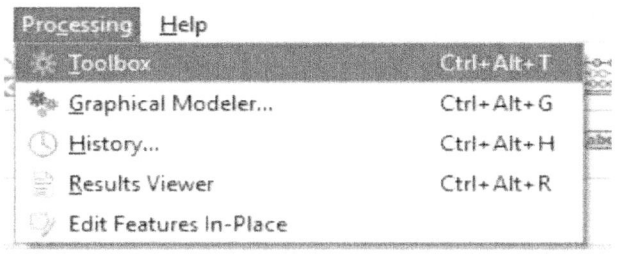

Once you've opened the processing toolbox you will see a list of categories and each category will have several analysis tools. Many third-party tools are also compatible with QGIS. These include SAGA, GRASS, and GDAL. Other tools can also be added as QGIS plugins.

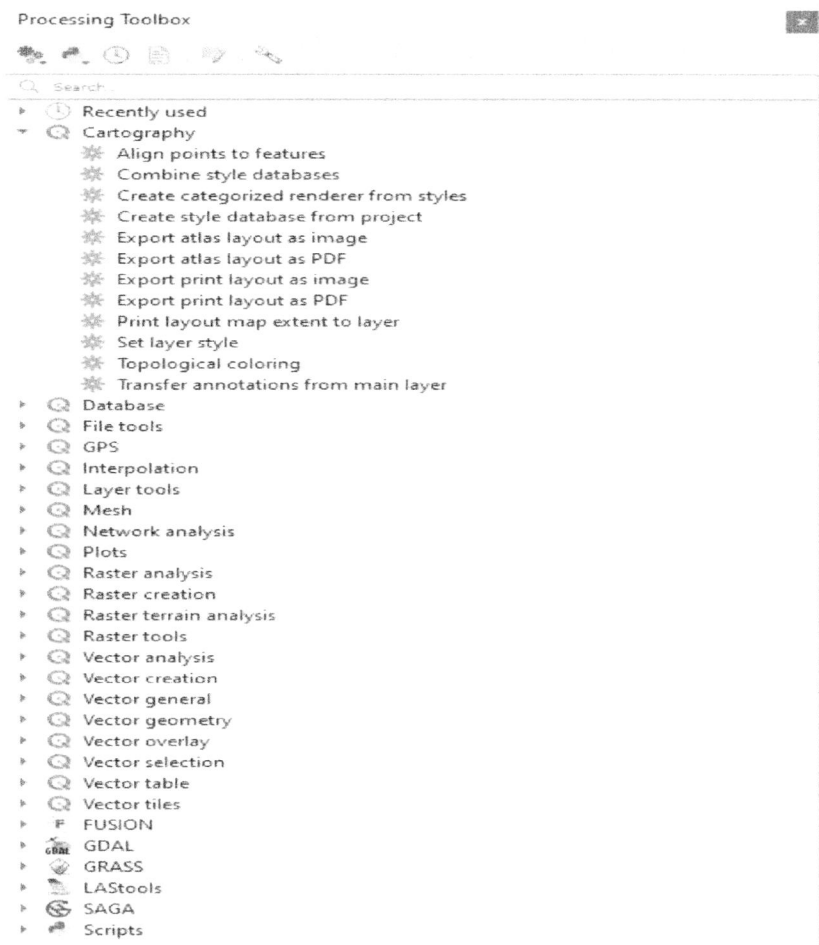

13. Create a Map Layout

Create a Map Layout

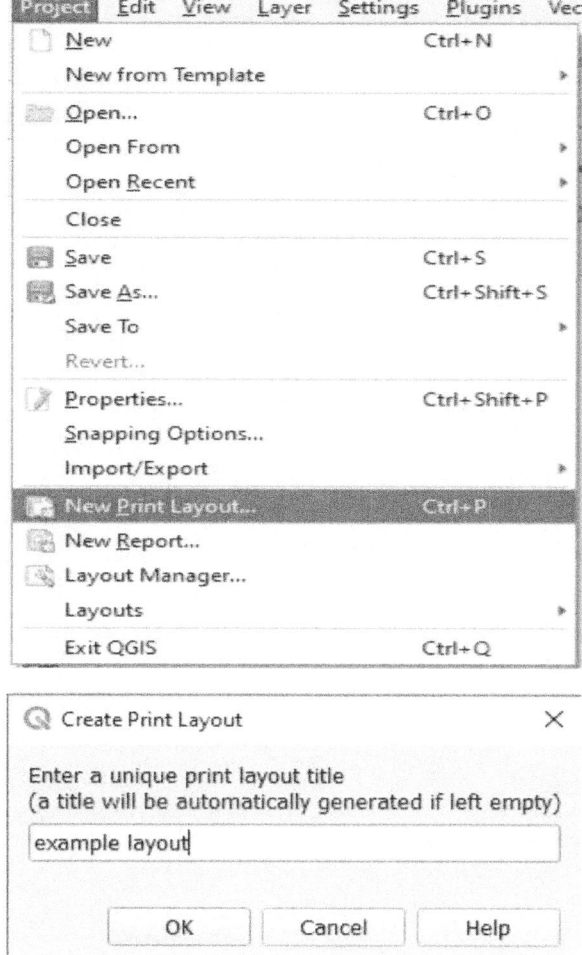

The QGIS Layout View

The QGIS Layout View works much like the main QGIS window. It contains its own toolbars and panels. If you do not see a toolbar or panel it can be added to the layout view by activating it from the 'View' option on the Main Menu in the same manner that we activated panels and toolbars in the main QGIS window.

The image below shows the default layout for the QGIS layout view. If your default view is different you can add panels and toolbars as described above. The most commonly used tools are labeled in the image. The 'Items' panel on the right side of the window is used to set properties for most items that are added to the map layout.

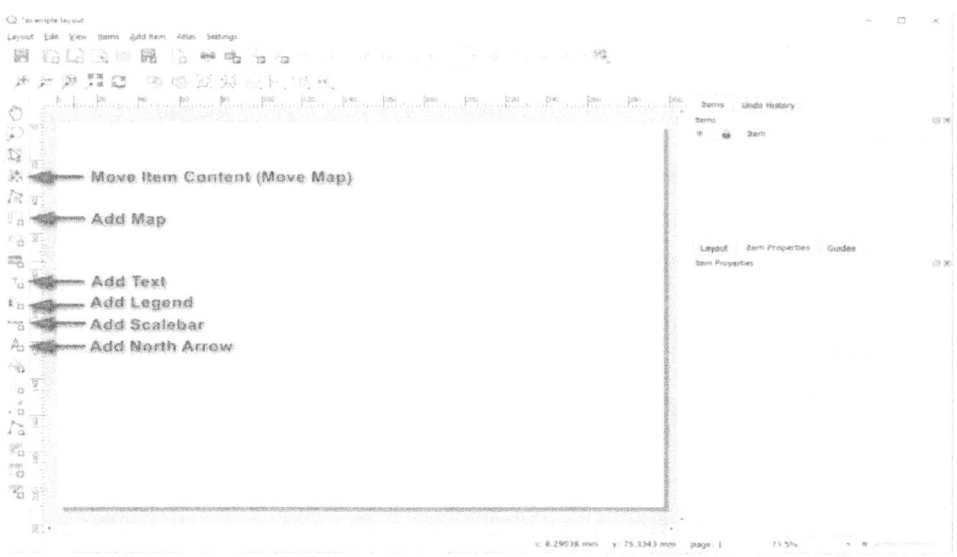

Edit layout size by right-clicking and selecting page properties. You can then adjust page properties in the panel on the right side of the screen.

Add Map to Layout

To add a map to the layout view, click the 'Add Map; button on the Toolbox toolbar (see image above). Then click and drag over the portion of the map layout where you want the map to appear. After you release from the drag the map will begin to load. If you are displaying many layers or a layer with many features it may take a little while for the map to load in layout view.

The layout view map will display layers the same way they are displayed in the main QGIS map canvas. In the 'Items' panel (right side of the screen) in layout view you will see options for adjusting the settings of the map in layout view. Once the map is symbolized the way you like you may wish to click 'Lock layers' and 'Lock styles for layers' in the Item Properties tab for the map item. This will ensure that your layout map stays the same even if you make changes in the QGIS map canvas.

To move the map around inside of the content window in layout view select the 'Move Item Content' button from the toolbar (see image above). Once this button is selected, click and drag the map to move it around inside its content window.

In layout view, the zoom and pan tools will zoom in and out on the layout/page, not on the map features. To zoom in on the map features make sure the 'Move Item Content' button is selected then hold control and use the mouse wheel to zoom in and out on the map. You can also adjust the scale in the Item Properties tab to zoom in or out.

Add Map Elements to the Layout

I'll explain how to add common elements to the map layout. For each of these elements, there are many additional settings and configurations that I do not have space to describe in full. The settings for each element can be accessed by selecting the element in the Items panel and changing the settings under the Item Properties tab.

Legend. Select the 'Add Legend' button then click and drag over the area you want the legend to appear. By default he a legend will be created for all active layers in the QGIS main window. When a layer is deactivated it will be removed from the legend (and from the map unless 'Lock layers' is selected as described above. You can change this behavior by unchecking 'Auto Update' In the Item Properties tab. You also need to turn off Auto Update if you want to change layer names from what appears in the QGIS main window.

Scale bar. Select the 'Add Scale Bar' button then click and drag over the area you want the scale bar to appear. You can adjust the scale bar appearance in the Item Properties tab.

North arrow. Select the 'Add North Arrow' button then click and drag over the area you want the north arrow to appear. You can adjust the north arrow symbol and appearance in the Item Properties tab.

Text and titles. Select the 'Add Label' (Add Text) button and drag over the area you want the text to appear. Then enter the text in the text box that will appear in the Item Properties tab. From the Item Properties tab you can also change the font style, size, and other settings.

Export Map

Once your map is complete you can export it as an image (recommended), PDF, or SVG file. Exporting is easy. Simply click on the icon for your desired export (or select the export option under the Layout tab on the main menu), specify a file name and format, then click 'Save'.

You will then be directed to set the export resolution and given an option to 'Crop to Content'. Crop to Content will limit the export to the area where objects actually appear on the layout plus the specified margin (if any). Then click 'Save' again.

Depending on the size of your map layout, the export may take a while. When it is complete you will be notified in the layout view.

Below is a quick example map I generated with the state outline data.

Error! Filename not specified.

14. QGIS Plugins

The QGIS Plugin repository hosts hundreds of plugins that extend the functionality of QGIS. The plugin repository can be accessed directly from QGIS by selecting Plugins > Manage and Install Plugins from the Main Menu.

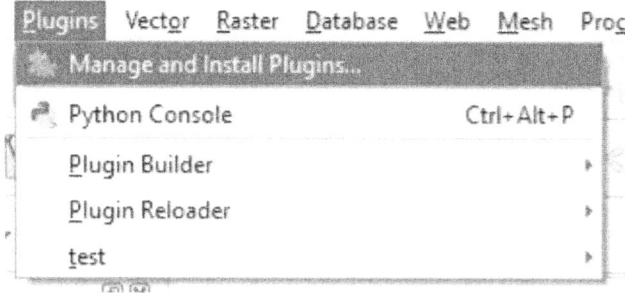

This will open a new window where you can browse and install plugins. For information on creating your own QGIS plugins with python,

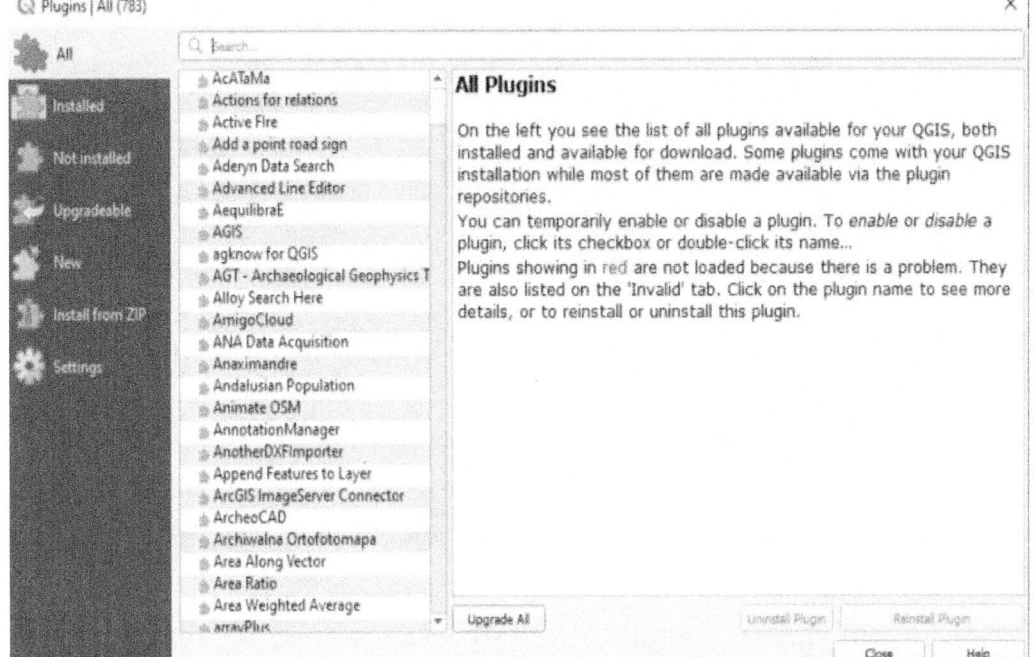

REFERENCES

Jump up to:[a][b] "QGIS Official Website". QGIS. *Retrieved 16 October 2013.*

QGIS related repositories on GitHub - NSA". GitHub. *Retrieved 31 March 2020.*

QuantumGIS (QGIS) – freie GIS-Software". Land Vorarlberg. *Retrieved 12 October 2013.*

QGIS related repositories on GitHub - LINZ". GitHub. *Retrieved 31 March 2020.*

Konrad Hafen, OpenSourceOptions, https://opensourceoptions.com/blog/qgis-tutorial-for-beginners/, 2022

Jump up to:[a][b][c] Cavallini, Paolo (August 2007). "Free GIS desktop and analyses: QuantumGIS, the easy way". The Global Geospatial Magazine.

OSGeo (February 2008). "OSGeo Annual Report 2007".

Gray, James (2008-03-26). "Getting Started With Quantum GIS". Linux Journal.

Tim Sutton (January 23, 2009). "Announcing the release of QGIS 1.0 'Kore'". *Retrieved 2009-01-26.*

QGIS for Android". Archived from the original on 21 October 2011. *Retrieved 25 September 2014.*

Download QGIS". QGIS.org. *Retrieved March 31, 2017.*

Commercial support". www.qgis.org. *Retrieved 2017-05-01.*

Visual Changelogs". www.qgis.org. *Retrieved 2022-07-16*

Releases - qgis/QGIS". *Retrieved 20 July 2022* – via GitHub.

QGIS. (2022, November 2). In Wikipedia. https://en.wikipedia.org/wiki/QGIS

Project details for Quantum GIS - Quantum GIS 0.9.0". Freshmeat. *Retrieved 2008-12-31.*

Changelog for QGIS 2.0 - Quantum GIS is now known only as 'QGIS'". QGIS. *Retrieved 1 January 2020*.

Printed in Dunstable, United Kingdom

76497854R00031